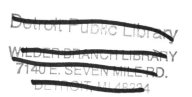
DEC 2005

WI

Amphibians

Sally Morgan

Chicago, Illinois

For information, address the publisher:
Raintree, 100 N. LaSalle, Suite 1200, Chicago, IL 60602

Produced for Raintree by
White-Thomson Publishing Ltd.

Consultant: Dr. Rod Preston-Mafham
Page layout by Tim Mayer
Photo research by Sally Morgan

Originated by Dot Gradations Ltd.
Printed in China by WKT Company Limited

09 08 07 06 05
10 9 8 7 6 5 4 3 2 1

Library of Congress Cataloging-in-Publication Data
Morgan, Sally.
   Amphibians / Sally Morgan.
      v. cm. -- (Animal kingdom)
   Includes bibliographical references (p.   ).
   ISBN 1-4109-1046-6 (lib. bdg. : hardcover) -- ISBN 1-4109-1342-2 (pbk.)
   1. Amphibians--Juvenile literature. [1. Amphibians.] I. Title.
II. Series: Morgan, Sally. Animal kingdom.
   QL644.2.M64 2004
   597.8--dc22
                                    2003026272

Acknowledgments
The publisher would like to thank the following for permission to reproduce copyright materials: Corbis pp.**9** bottom (Joe McDonald), **21** top and **25** top (Chris Mattison/Frank Lane); Digital Vision Title page, contents page, pp.**4**, **6**, **18**, **36**, **42**, **44** bottom, **45**, **46**, **48**; Ecoscene pp. **31** top (Kjell Sandved), **33** (John Pitcher), **38** (Anthony Cooper), **40** (R.A. Beatty), **43** (Anthony Cooper); Ecoscene/Papilio pp.**7** top, **10** top and **11** (Robert Pickett), **15** top (Paul Franklin), **15** bottom (David Manning), **17** top (Robert Pickett), **19** top and **24** (Jamie Harron), **26–27** and **27** top (Paul Franklin), **37** bottom (Robert Pickett), **41** bottom (Paul Franklin); Nature pp.**8** (Fabio Liverani), **9** top (Phil Savoie), **10** bottom (Reijo Jurrinen), **13** (Mark Payne Gill), **17** bottom (Ingo Arndt), **23** bottom (Barry Mansell), **26** top (Fabio Liverani), **28** (Morley Read), **32** top (John Downer Productions), **39** top (Jose B Ruiz), **41** top (George McCarthy), **44** top (John Cancalosi); NHPA pp.**5** top (T. Kitchin and V. Hurst), **7** bottom (Daniel Heuclin), **12** top (ANT), **12** bottom (Nobert Wu), **14** (Stephen Dalton), **16** (Kevin Schafer), **18** bottom (Daniel Heuclin), **20** (Stephen Dalton), **21** bottom (Kevin Schafer), **22** (Robert Erwin), **23** top, **25** bottom and **29** top (Daniel Heuclin), **28** bottom (Jenny Sauvanet), **30** (Daniel Heuclin), **32** bottom (Stephen Dalton), **34** (Pierre Petit), **35** top (Anthony Bannister), **35** bottom (Daniel Heuclin), **36** top (James Carmichael), **39** bottom (Bill Coster); Photodisc pp.**5** bottom, **31** bottom.

Cover photograph of strawberry poison frogs reproduced with the permission of NHPA (T. Kitchin and V. Hurst).

Every effort has been made to contact copyright holders of any material reproduced in this book. Any omissions will be rectified in subsequent printings if notice is given to the publisher.

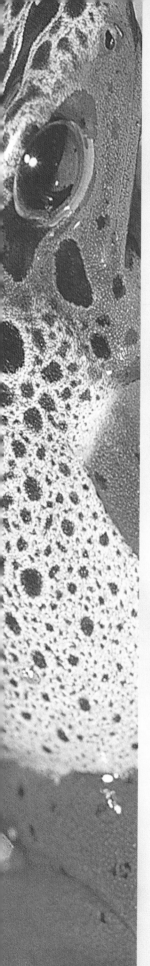

Contents

Introducing Amphibians

Frogs, toads, salamanders, and newts are all amphibians. Most amphibians live part of their lives on land and part in water. However, some amphibians spend their entire lives in water or on land.

Amphibians belong to a large group of animals called vertebrates. These are animals with backbones. Other vertebrates include fish, reptiles, birds, and mammals.

Breathing

Although most amphibians have lungs, many rely on their skin to breathe. These amphibians have a smooth, moist skin, which allows the passage of oxygen into the body.

Classification key	
KINGDOM	Animalia
PHYLUM	Chordata
SUBPHYLUM	Vertebrata
CLASS	**Amphibia**
ORDERS	3-Caudata, Gymnophiona, Anura
FAMILIES	44
SPECIES	5,565 and increasing

▼ All amphibians, including this ornate horned toad, undergo metamorphosis.

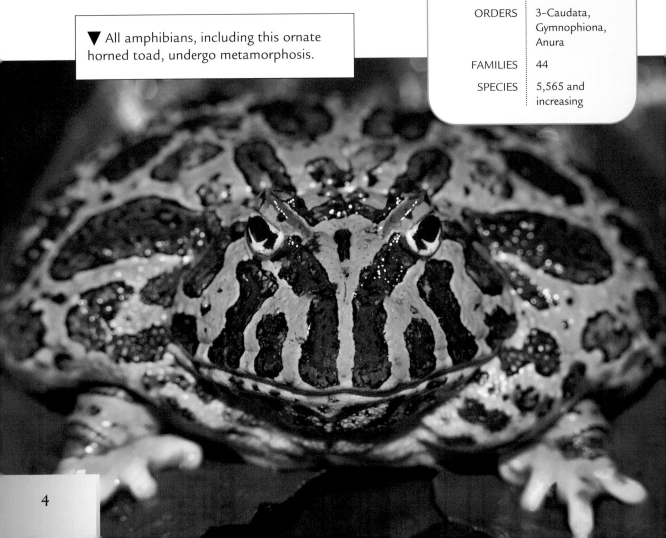

Cold-blooded animals

Amphibians are ectothermic, or cold-blooded. This means that their body temperature is similar to that of the environment around them. If the external temperature falls, an amphibian's body temperature falls, too. At low temperatures, amphibians cannot produce enough body heat to stay active, so they become dormant, or inactive, until temperatures rise.

Metamorphosis

One of the most important features of amphibians is the fact that they undergo a change in appearance called metamorphosis. Frogs, for example, start life as tadpoles.

▼ This blotched tiger salamander is a tailed amphibian. Salamanders are found mostly in the northern hemisphere.

Smooth, moist skin allows oxygen to pass into the body.

▼ Frogs are tailless amphibians with four long, muscular legs and webbed hind feet.

Long, muscular hind legs for hopping and leaping.

Classification

Living organisms are classified, or organized, according to how closely related one organism is to another. The basic group in classification is the species. For example, human beings belong to the species *Homo sapiens*. A species is a group of individuals that are similar to each other and that can interbreed with one another. Species are grouped together into genera (singular: genus). A genus may contain a number of species that share some features. *Homo* is the human genus. Genera are grouped together in families; families are grouped into orders; and orders are grouped into classes. Amphibians are in the class Amphibia. Classes are grouped together in phyla (singular: phylum) and finally, phyla are grouped into kingdoms. Kingdoms are the largest groups. Amphibians belong to the animal kingdom.

Webbed hind feet propel the frog easily through water.

Amphibians are found in freshwater habitats on every continent except Antarctica. They cannot live in salt water. Most live near water because they have to return to water in order to breed. However, many salamanders live a completely terrestrial life and lay their eggs on land.

Moist skin

Amphibians have smooth, thin skin with no scales or hair. They can breathe through their skin, absorbing oxygen from either air or water. For this to happen the skin must be kept moist at all times. Amphibians produce a mucus that makes their skin moist and slippery to the touch. Although amphibians have lungs, they breathe mostly through their skin. They use their lungs when they are active and need more oxygen. In water, adult amphibians come up to the surface of the water to take air into their lungs.

Larval amphibians, or those at an early stage of development such as tadpoles, breathe with gills. Gills have a large surface area in contact with the water, which means that more oxygen can be picked up by the blood flowing through them. In a young tadpole, the gills are outside the body, but as the tadpole becomes older the gills are covered over. Eventually they are replaced by lungs. These older larvae rely mostly on their lungs to obtain oxygen, although a small amount of oxygen will pass through the skin. Some amphibians that live in water permanently, such as the axolotl, breathe mostly through their gills.

▼ This tinkered frog has a smooth skin that must be kept moist to allow oxygen to enter its body.

▲ When toads rest at the surface, their eyes lie above the water, allowing them to watch out for predators.

Moving around

Animals that live in water must be able to swim. Adult frogs and toads, for example, have long legs with webbed feet that push them through the water. However, on land frogs can use their powerful legs for crawling, hopping, and leaping. Newts have a long tail with a fin that they use to swim and steer. The eyes of a frog have vertical pupils. They allow the frog to see above the surface of water while keeping the rest of its body under water.

▶ Newts, such as this palmate newt, are found in water during the breeding season. They obtain oxygen mostly through the skin and mouth.

Life Cycle

Amphibians are the only vertebrates to undergo metamorphosis, or a complete change in their appearance.

Egg laying

In most amphibian species, the female lays her eggs in water and they are fertilized by the males. The eggs may be laid singly or in batches. The batches of eggs may form long chains or clusters. Amphibian eggs are different from those of reptiles and birds because they are not protected by a shell. Each egg is made up of a small black spot surrounded by a jelly coating. The black spot is the fertilized egg—a tiny embryo with a much larger yolk sac beneath it. The embryo has all the food it needs in the yolk sac. Some salamanders live on land and do not return to water to breed. They lay a small number of large eggs and the larvae develop inside the eggs before hatching as miniature adults.

Larval amphibians

The embryo grows inside the egg, gradually getting larger and more elongated in shape. The jelly starts to break down and the egg hatches. A newly hatched amphibian is called a larva, or, in the case of frogs and toads, a tadpole. The larvae feed constantly. Often the first food they eat is the remains of the jelly.

◀ Toads, such as this green toad, lay their eggs in long chains that they wrap around underwater plants so they do not drift away.

▲ This glass frog is guarding her clutch of eggs on a leaf.

The tadpoles of frogs and toads and the larvae of newts feed on plants, either filtering their food from the water or scraping off plant material. After a few weeks, they change their diets and become carnivorous. Salamander larvae are often predatory. As the larvae grow, they begin to undergo a gradual metamorphosis, becoming more like an adult. For example, frog and toad tadpoles grow legs and their tails disappear. The larvae of newts and salamanders already look like the adult, so they do not have to undergo such a dramatic metamorphosis. The shape of the tail fin changes and their skin becomes thicker.

Viviparous salamanders

A few salamanders do not lay any eggs. Instead the eggs are kept inside the body of the female. The embryos develop within the eggs, using nutrients supplied by the female. The female gives birth to a small number of well-developed larvae. This is called vivipary, or live birth.

Amazing facts

● Fire salamanders are viviparous. The faster-growing young of fire salamanders may eat their smaller siblings while still inside their mother's body.

● The tadpoles of the Surinam horned toad are very aggressive. They are carnivorous from the moment they hatch. They will attack tadpoles of other species and even each other.

▲ This female Jefferson salamander has laid a large clutch of eggs. The eggs will hatch into larvae.

The Common Frog

In early spring, common frogs (*Rana temporalis*) travel to ponds to breed. They often gather in large numbers. Each female lays her eggs, which are then fertilized by the males.

Frogspawn

The female frog lays a clutch, or cluster, of about 100 to 200 eggs. The clusters are known as frogspawn. Each newly laid egg consists of a blob of protective jelly surrounding a black dot. The black dot is the embryo that grows into a tadpole. The cells that form the embryo divide and increase in number. Soon the blob becomes longer, and it is possible to make out a tiny tadpole. After about ten days, the protective jelly turns to liquid so the tadpole can move around. It wiggles its way out of the jelly and into the water.

Tadpoles

The young tadpoles are herbivores, or plant eaters. They feed on pondweed and microscopic algae in the water. The tadpoles grow quickly, and their long tails help them to swim around.

▲ At first tadpoles have only a head and a tail. The legs appear later.

▼ Male and female frogs gather in breeding ponds in early spring.

Each tadpole has tiny external gills on either side of the head. They use these gills to breathe by taking in oxygen from the water. These external gills become smaller, and after four weeks they disappear completely. They are replaced by internal gills, which are protected on the outside by a flap of skin.

Two small bulges appear at the back of the body on either side of the tail. The bulges develop into legs, each of which ends in a webbed foot. The front legs appear a few weeks later. The diet of the tadpole changes from plants to small animals.

There are internal changes, too. The lungs begin to grow and they gradually take over from the internal gills. When this happens, the tadpoles come to the surface of the water to take gulps of air.

Classification key

CLASS	Amphibia
ORDER	Anura
FAMILY	Ranidae
GENUS	*Rana*
SPECIES	***Rana temporalis***

Final changes

Finally, the shape of the tadpole changes. The tail becomes shorter until just a stump remains. The backbone becomes more obvious. The round mouth of the tadpole changes to the much wider mouth of the frog, and the eyes stick out more. The complete metamorphosis takes about twelve weeks. The tiny frogs leave the water and live on land for the next few years until they are fully grown and ready to breed.

Amazing fact

○ The paradoxical frog gets its strange name from the fact that its tadpoles live for a long time and they grow to four times the length of the adult frog. The word *paradoxical* means "absurd" or "contradictory."

▲ Once a tadpole has developed its front legs, the tail becomes shorter and shorter.

11

Extreme Survivors

Most amphibians live in damp places and need water to survive. However, some amphibians can be found in extreme environments such as hot deserts or the freezing Arctic.

Surviving drought

Deserts are dry places where one might not expect to see any amphibians. The daytime temperatures are high, and rain is infrequent. When it does rain, there are heavy downpours. However, a few frogs and toads have adapted to this dry habitat. They tend to be active at night when the temperatures decrease. Desert frogs and toads survive long periods of dry weather by burrowing underground. They produce a mucus that they spread over their bodies. The mucus hardens to prevent water from escaping. Once it is safely cocooned in mucus, the frog goes into a state of torpor, or becomes completely inactive. It may stay like this for several months or even years until the rains return.

Desert frogs and toads are able to lose lots of body water. They can also take up water quickly. The spadefoot toad can absorb water simply by sitting on a damp patch of ground. The skin touching the ground is thin and rich in blood vessels and can quickly absorb the water. Some desert frogs and toads can store water in their bodies. The water-holding frogs have large, baggy glands under their skin that can swell up with water. As much as half their body weight can be water.

▲ Burrowing frogs store water in their bladders. This allows them to stay underground without drying out.

◀ A California newt can survive fire by secreting mucus over its body.

▲ This male bullfrog is digging a channel to allow his tadpoles to escape to a larger pool before they dry out.

Arctic frogs

Amazingly, frogs can be found in the far north at the edge of the Arctic. Here they have to survive long, cold winters. They do this by creeping into a safe place where they go into torpor until the temperature rises again.

Fire survivors

Fires are common in the grasslands and forests of California. California newts have adapted to this danger. They secrete mucus all over their bodies. The mucus protects them so well that they can walk right through the flames.

Amazing facts

- The crab-eating frog of Southeast Asia lives beside slightly salty water, where it is said to feed occasionally on crabs.
- The Aboriginal people of Australia use water-holding frogs as a source of water.
- Pools of water do not last long in deserts. The eggs of the spadefoot toad hatch in three days and the tadpoles complete their metamorphosis in six to eight days.

Color

Amphibians display an amazing range of colors, from red and black spots to vivid blues and greens. The color of a frog's skin comes from pigments in the skin cells.

Color change

Many amphibians can change the color of their skin by making the area of pigment larger or smaller. This process takes several minutes. The color of a frog's skin can alter the amount of heat its body absorbs. Dark colors absorb more heat than light colors. By making their skin lighter, some amphibians can reflect heat away from their bodies, and this allows them to cool down. If the amphibians are too cool, they can increase the amount of heat they absorb by making their skin darker.

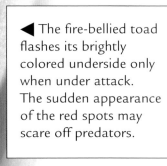
◄ The fire-bellied toad flashes its brightly colored underside only when under attack. The sudden appearance of the red spots may scare off predators.

▲ This narrow-headed frog is so well camouflaged it can barely be seen against the background of tree bark.

Amazing facts

○ To scare off predators, the Italian spectacled salamander raises its tail over its back in a defensive position to reveal the brightly colored underside of its tail.

○ The Corroboree frog is a vivid yellow with black stripes. The reason for its bright colors is uncertain because it does not produce any poisons from its skin.

► The yellow spots of the fire salamander are a warning that this amphibian secretes poisons.

Color for camouflage

Some amphibians use color to blend in with the background and become camouflaged. Some form a pattern of colors that breaks up the outline of the animal so that predators cannot see them resting on the ground. Many frogs and toads that live on forest floors look like dead leaves. Some even have a line that looks like the middle vein of a leaf running down their backs.

Warning colors

Other amphibians, such as the fire salamander and poison frogs, are brightly colored to make them stand out. These are warning colors that tell would-be predators that they are poisonous and not to eat them. Others use color to surprise their predators. For example, when the fire-bellied toad is attacked, it flips over and reveals a brightly colored underside to scare away its attacker. There are some brightly colored frogs and toads that are not poisonous. They have copied the warning colors of a poisonous species to trick predators into avoiding them.

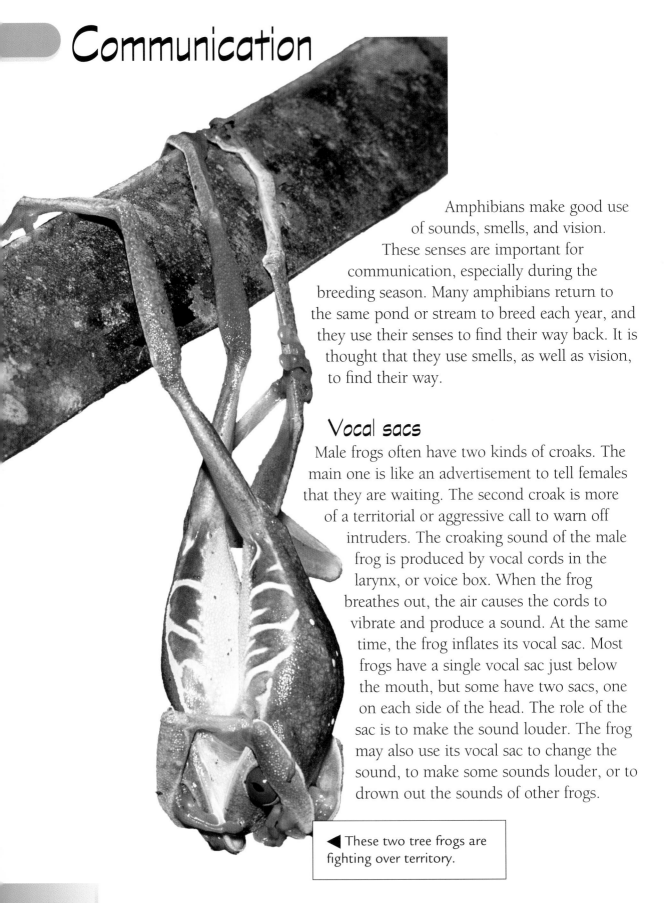

Communication

Amphibians make good use of sounds, smells, and vision. These senses are important for communication, especially during the breeding season. Many amphibians return to the same pond or stream to breed each year, and they use their senses to find their way back. It is thought that they use smells, as well as vision, to find their way.

Vocal sacs

Male frogs often have two kinds of croaks. The main one is like an advertisement to tell females that they are waiting. The second croak is more of a territorial or aggressive call to warn off intruders. The croaking sound of the male frog is produced by vocal cords in the larynx, or voice box. When the frog breathes out, the air causes the cords to vibrate and produce a sound. At the same time, the frog inflates its vocal sac. Most frogs have a single vocal sac just below the mouth, but some have two sacs, one on each side of the head. The role of the sac is to make the sound louder. The frog may also use its vocal sac to change the sound, to make some sounds louder, or to drown out the sounds of other frogs.

◀ These two tree frogs are fighting over territory.

Croaking frogs

A lot of croaking takes place during the breeding season. Sometimes male frogs gather in one place to breed. There is a short period of intense activity during a few nights when all the frogs croak in order to attract the attention of as many females as possible. Because the sound of many frogs croaking is louder than just one frog croaking, each male has a greater chance of finding a female. Some species of frogs gather in traditional breeding places, but they do not call at the same time. Each night only a few males will croak. This means that the breeding goes on for a longer period of time, sometimes for many months.

Male frogs will also use sound in disputes with other males. When another male comes into their territory they will make a sound that warns the other male. If the other male ignores the warning, the males may fight.

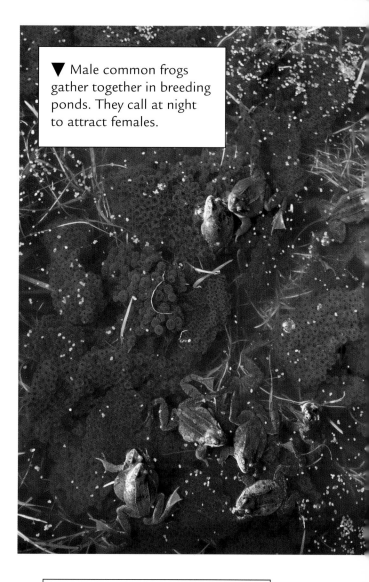

▼ Male common frogs gather together in breeding ponds. They call at night to attract females.

▼ The natterjack toad has a single vocal sac below its throat that inflates when it croaks.

Amazing facts

- The male Tungara frog calls more than 7,000 times in a single night. It attracts females as well as bats. The bats swoop down and grab the frogs.

- Sometimes male frogs make calls to interfere with a neighboring frog that is trying to attract females!

Amphibian Orders

There are 5,565 amphibian species, of which 4,883 are frogs and toads, 517 are newts and salamanders, and 165 are caecilians. The number of species of amphibians is rising, unlike that of other vertebrates. Since 1985, the total number of recognized species has increased by more than one third, and the number is still growing as new species are discovered. However, many species are under threat of extinction, so the numbers may start to fall again.

Classifying amphibians

Amphibians are placed in the class of Amphibia. Within this class there are three orders—Anura (frogs and toads), Caudata (newts and salamanders), and Gymnophiona (caecilians).

When classifying amphibians, biologists look at features such as the number of vertebrae, which are the small bones that make up the backbone. For example, salamanders have a long, flexible backbone with many vertebrae, while frogs and toads have a short, strong backbone with far fewer vertebrae. Most amphibians have four limbs. The front limbs have four toes while the hind limbs have five toes. However, some salamanders either have no hind limbs or have fewer toes, and the caecilians are completely limbless. Frogs and toads have much larger heads than the other amphibians. However, most amphibians have well-developed eyes that they use to find and catch food.

◄ Tree frogs, which have no tails, belong to the order Anura.

Fossil amphibians

Amphibians were among the first vertebrates to live on land more than 300 million years ago. Scientists think they evolved from fish that had fleshy front limbs and could flip from pool to pool over land. The earliest fossil amphibians are called labyrinthodonts. They were large animals more than 3 feet long that walked very slowly. Most of these early amphibians died out, but two groups survived. One group was Lissamphibia, which became the modern amphibians that we see today. The other group evolved into reptiles.

Amazing facts

- Miniature salamanders, less than an inch long, are among the smallest of all vertebrates. These salamanders have tongues that stretch more than half the length of their bodies. They use their tongues to catch insects.

- Chinese scientists have discovered Asia's oldest frog fossils. One is nearly 125 million years old.

- The backbone of a frog consists of approximately twelve bones. A salamander may have as many as 100 bones while a caecilian has up to 250.

▲ The marbled newt, with its mottled black-and-green skin, is more brightly colored than other European newts.

19

Tailed Amphibians

Salamanders, newts, and their relatives belong to the order Caudata. They are found throughout North America, Europe, and parts of Asia, but they are not found in the southern hemisphere. They have long, flexible bodies and tails, and four relatively short but sturdy limbs. Some tailed amphibians do not have hind limbs. This order of amphibians is thought to be most similar to the earliest amphibians that have long been extinct.

Newts and salamanders look very similar. The word *salamander* is used to describe a tailed terrestrial, or land-dwelling, amphibian. The word *newt* comes from the Anglo-Saxon word *efete* or *evete,* and it refers to those tailed amphibians that return to water to breed each spring. Despite these words, some salamanders are found in water and newts spend many months on land.

Smooth skin

Newts and salamanders have smooth, flexible skin with no scales. Although they have lungs, these amphibians use their skin to obtain oxygen. This can happen only if their skin is kept moist. This means that the newts and salamanders have to stay in damp or wet habitats, such as bogs, marshes, and wetlands.

Some tailed amphibians are completely aquatic. They spend their entire lives in water. Many of these aquatic amphibians have kept their external gills, which they use to breathe.

▼ The smooth newt spends several months each year in water.

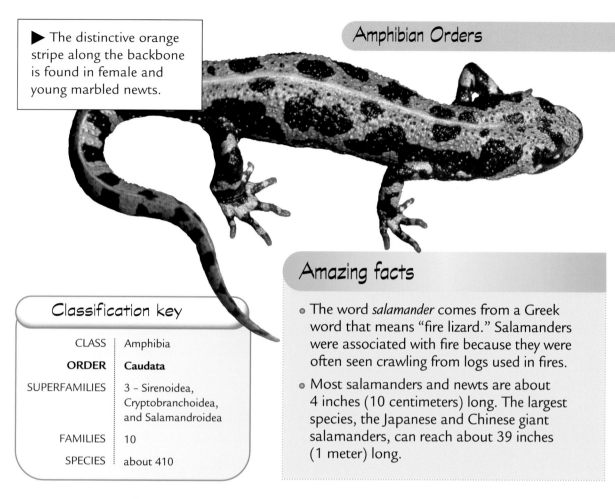

▶ The distinctive orange stripe along the backbone is found in female and young marbled newts.

Classification key

CLASS	Amphibia
ORDER	**Caudata**
SUPERFAMILIES	3 – Sirenoidea, Cryptobranchoidea, and Salamandroidea
FAMILIES	10
SPECIES	about 410

Amazing facts

- The word *salamander* comes from a Greek word that means "fire lizard." Salamanders were associated with fire because they were often seen crawling from logs used in fires.

- Most salamanders and newts are about 4 inches (10 centimeters) long. The largest species, the Japanese and Chinese giant salamanders, can reach about 39 inches (1 meter) long.

Life cycle

Not all newts and salamanders have the same life cycle. Some live entirely in water and never go onto land. Others return to water to breed. Their larvae live in water while they undergo metamorphosis. Some tailed amphibians live only on land and lay eggs that hatch into miniature adults. Others give birth to live young.

▲ The axolotl looks like a giant tadpole with external gills. The adults keep their external gills because they are completely aquatic.

Three superfamilies

Within the order of Caudata there are three superfamilies. The largest is Salamandroidea, which includes newts, European salamanders, and mole salamanders. The superfamily Sirenoidea contains sirens, while the superfamily Cryptobranchoidea contains giant salamanders and Asiatic salamanders.

Sirens and Giant Salamanders

Sirens and giant salamanders belong to the two superfamilies of Sirenoidea and Cryptobranchoidea.

Sirens

There are four species of siren, and they are found only in some southern states of the United States and northern Mexico. They live in shallow water in ditches, streams, and lakes. Sirens are eel-like animals, ranging in length from 4 to 35 inches (10 to 90 centimeters). They spend their entire lives in water. Sirens have external gills, which they use to breathe in water. They have a long body and no hind limbs. Their front limbs are small and are found just behind the gills. Their eyes are very small and have no eyelids. Sirens are usually brownish-green with pale spots.

All sirens are meat eaters, or carnivores, that feed on small animals such as crayfish, worms, and snails. A siren feeds by sucking mud that contains these animals into its mouth. Unlike most amphibians, sirens do not have teeth at the front of the mouth. They have horny beaks instead. Sirens can survive dry periods by burrowing into the mud and covering themselves in mucus that forms a protective covering, or cocoon, around them. They go into a state of torpor and can survive like this for months.

Classification key

CLASS	Amphibia
ORDER	Caudata
SUPERFAMILY	**Sirenoidea and Cryptobranchoidea**
FAMILIES	3
SPECIES	42, possibly more

▼ Sirens have long, eel-like bodies and external gills.

▼ Hellbenders are giant North American salamanders that have heavy heads and four short legs.

Amazing facts

- The greater siren can survive encased in its cocoon for two years without food.
- In Japan giant salamanders are caught and eaten because they are considered a delicacy.
- One giant salamander in captivity lived to 52 years of age.

Giant salamanders

Giant salamanders are about 35 inches (90 centimeters) long. They live in rivers and streams and are aquatic. However, unlike the sirens, they do not have external gills. They breathe through their skin. There is a fold of skin along the sides of their bodies that increases the surface area over which oxygen can enter the body. They use their lungs, too, and they come to the surface to take gulps of air. Giant salamanders are nocturnal and come out at night to hunt. They eat a variety of animals that they find in the water, including fish, insects, and snails.

▼ The Chinese giant salamander lives in fast-flowing mountain streams where the water has plenty of oxygen.

Salamanders and Newts

Most tailed amphibians belong to the superfamily Salamandroidea, which includes European salamanders and newts, mole salamanders, olm, mudpuppies, torrent salamanders, and lungless salamanders.

Newts

Newts are true amphibians. Adult newts spend up to half the year in water and the rest of the year on land. Their body undergoes a change when they return to the water. Their skin becomes smooth and able to absorb oxygen, their tails become more streamlined to help them swim, and their eyes change shape to help them focus under water. Some newts develop webbed feet. When they are in water, newts can breathe in one of three ways: through their skin, through the lining of their mouths, or through their lungs. When they are more active, newts cannot get enough oxygen through their skin and mouth, so they come up to the surface to gulp air. This can be risky because predators, such as herons, may spot them.

In spring, the mature adults travel to their breeding ponds, where they lay eggs in water. Their larvae grow rapidly and develop legs. By the end of summer they leave the water looking like small adults. The larvae spend the next few years on land before returning to the ponds to breed.

Classification key

CLASS	Amphibia
ORDER	Caudata
SUPERFAMILY	**Salamandroidea**
FAMILIES	7
SPECIES	approximately 370

▼ Newts usually return to the same pond to breed year after year, often traveling several miles.

▶ The tiger salamander is one of the most widely distributed of the mole salamanders.

Mudpuppies and waterdogs

These salamanders are totally aquatic, and they have feathery external gills as well as lungs. They live in ponds, lakes, and streams. They are predators that feed on small aquatic animals. Females lays their eggs in spring, sticking each one to an underwater rock or log. The male guards the eggs until they hatch five to nine weeks later.

Mole salamanders

Mole salamanders are given this name because they live in a burrow for much of their lives. They are rarely seen except during the breeding season when they travel to ponds to breed. They have heavy bodies with smooth, shiny skin that is often brightly colored. The spotted salamander, for example, is a mole salamander.

▼ The olm has adapted to life in dark caves. It has a long, thin body with pink skin, two pairs of tiny limbs, and external gills.

Amazing facts

- California newts have skin secretions that are among the most poisonous substances known.
- The spiny newt has long, sharp, pointed ribs. If grabbed by a predator, the ribs push out through poison glands in the skin.

The Great Crested Newt

The great crested newt is a large European newt that spends up to five months of the year in ponds and lakes. The adult male is up to 5.5 inches (14 centimeters) long. Its orange or yellow underside with black blotches warns predators that it is poisonous.

Habitat

Great crested newts need several different types of habitats during their lives. They spend much of the year on land and can be found in woodlands and grassland, where they feed on earthworms, insects, spiders, and slugs. They are nocturnal and hide on land during the day in burrows or under logs, stones, and vegetation. They become inactive between October and late February, when they shelter under piles of leaves or logs, or in hollow tree stumps and stone walls.

Breeding colors

In early spring, the newts become active again and travel to their breeding ponds—often the ones in which they were born.

▲ Great crested newts are able to breed when they are about two to three years old.

Classification key

CLASS	Amphibia
ORDER	Caudata
SUPERFAMILY	Salamandroidea
FAMILY	Salamandridae
GENUS	*Triturus*
SPECIES	***Triturus cristatus***

The males undergo an impressive change in appearance when they develop their breeding colors. They have a large, jagged crest along the length of their backs, a blue-white streak down the sides of their tails and more obvious orange undersides. Females do not have a crest, and they have a yellow-orange stripe along the undersides of their tails.

▲ Newt larvae have well-developed tails and external gills.

Courtship and change

The newts undergo a complicated courtship. A male swims near a female with his crest standing up. He may nudge the female with his nose and swim around her. Egg laying lasts from March to mid-July. Although each female lays about 200 to 300 eggs, she lays only 2 or 3 eggs per day. These are carefully wrapped in the leaves of aquatic plants. The larvae hatch after about 3 weeks. Larval newts usually feed on tadpoles, worms, insects, and insect larvae. Their metamorphosis into air-breathing youngsters takes about four months. At this stage, newts are ready to leave the pond and live in damp habitats near water.

Amazing facts

- Great crested newts can live for 27 years, or possibly more.
- Females choose a mate according to the size of the crest.

◀ The adult great crested newt hunts in ponds for other newts, tadpoles, young frogs, worms, insect larvae, and water snails. It also hunts on land for insects, worms, and other invertebrates.

27

Caecilians

Caecilians have long bodies and no legs or tails. These unusual amphibians are the least studied because they spend so much time in burrows. Caecilians are found mostly in Central and South America, southern China, west Africa, and parts of Southeast Asia.

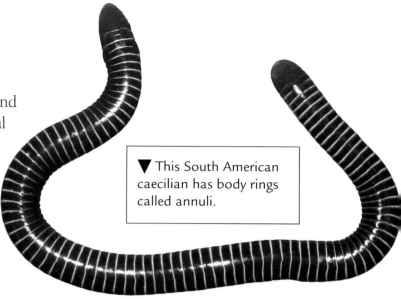

▼ This South American caecilian has body rings called annuli.

Bodies for burrowing

Caecilians are adapted to burrowing. They have fat and elongated bodies that look as if they are made up of lots of segments, like an earthworm. However, this is only an illusion. In fact, a caecelian's skin is joined to the bones underneath so it does not get ripped off or damaged during burrowing. Caecilians burrow by using their heads like shovels. For this reason, their skulls are particularly large, heavy, and bony. Their eyes are small as they spend most of their time underground and do not rely on sight. The land-dwelling caecilians burrow into soil, while the aquatic species burrow into mud under water. They move through the ground by contracting, or tightening, their muscles. As the muscles contract, they squeeze on the fluid inside the body, making the body get longer and move forward. One unusual feature of the caecilians is a small tentacle below each eye. The tentacle detects chemicals and is used to find prey.

▼ This caecilian is eating a worm. The body of the caecilian looks very similar to that of the worm.

Amazing facts

- Caecilians range in size from just 2.7 inches (7 centimeters) to 63 inches (1.6 meters) long.
- Aquatic caecilians look like eels. They have fins on their tails that they use to swim.
- Linnaeus's caecilian has a rear end covered in a hard shield. It also has up to 300 grooves in its skin, and within each groove the skin is covered in tiny scales.

Waiting for food

Caecilians are carnivores that feed on earthworms, termites, and other animals they find while burrowing through the soil. Sometimes caecilians emerge from their burrows and wait for prey to pass close by. This way they may catch lizards, grasshoppers, and crickets. However, while the caecilians are outside the burrow, predators such as snakes and birds may catch them.

Live birth

Most caecilians are viviparous and give birth to live young. Sometimes the young are kept within the female's body up to eleven months. A few caecilians lay their eggs in burrows near streams. After they hatch, the tiny larvae wiggle into the water. Egg-laying caecilians guard their eggs from predators until they hatch.

▼ A few caecilians are aquatic and spend their lives in water.

Classification key	
SUBCLASS	Amphibia
ORDER	**Gymnophiona**
FAMILIES	6
SPECIES	176

Frogs and Toads

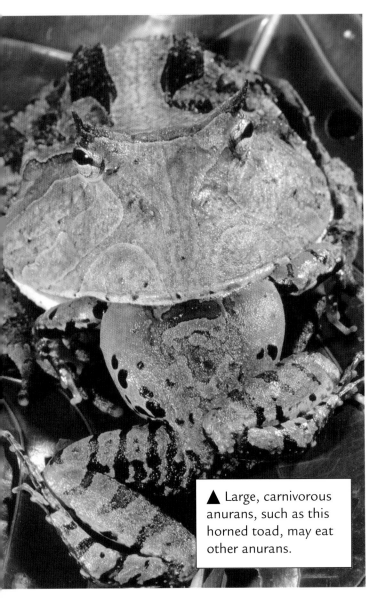

▲ Large, carnivorous anurans, such as this horned toad, may eat other anurans.

Frogs and toads make up most of the amphibians. They belong to the order Anura. Frogs and toads are found in a range of habitats from rain forests to mountain slopes. They are most common in the warmer parts of the world, especially the tropics, but they can be found as far north as the Arctic. Frogs and toads spend part of the year in water and part on land.

Anuran features

Adult frogs and toads can be distinguished from other amphibians because they have long hind limbs and no tails. The backbone is much shorter than in other amphibians. It is a stiff rod that gives support when the frog or toad is jumping and landing. Frogs and toads do not have necks, so they cannot move their heads from side to side. Most of the time they breathe through their skin and the lining of their mouths. When frogs and toads are active in water, they rise to the surface to gulp air into their lungs.

Amazing facts

- The African clawed toad has what looks like a line of white stitch marks along its sides. These contain sense organs that allow it to detect vibrations made by prey or predators in the water.
- The cane toad is the world's largest toad at 9.5 inches (24 centimeters) long. When threatened, it squirts poison into the eyes of its attacker.

▲ Anurans that spend a lot of time in water have webbed hind feet like this African clawed toad.

Feeding

Frogs and toads are carnivores that make good use of their senses, especially their sight, to find their prey. They catch prey using their long tongues, which are attached only at the front of the mouth. This allows the frog or toad to flick its tongue quickly to catch food. The upper surface is very sticky and can trap prey. A wide mouth allows frogs and toads to swallow large prey. Some of the larger frogs will hunt prey such as rats and snakes. To swallow, a frog or toad closes its mouth and eyelids and presses down on the food with its eyeballs.

Frog or toad?

Many people describe toads as having dry, warty skin and being heavier and slower moving than frogs. In reality, there is no difference. Strictly speaking, the word *toad* should be used only for members of the genus *Bufo* such as the common toad, *Bufo bufo*. The African clawed toad, *Xenopus laevis*, lives in water, has smooth skin, and is a frog, not a toad.

▼ Frogs and toads have large, bulging eyes with eyelids for protection from dust and soil. The eardrum can be seen on the side of the head.

Classification key

CLASS	Amphibia
ORDER	**Anura**
FAMILIES	28
SPECIES	approximately 4,750

In or out of water?

Frogs that spend a lot of their time in water, such as the edible frog and the bullfrog, have a very smooth skin and a streamlined body. Their hind limbs are long and muscular and they end in webbed feet. These features help them move through water with ease. They go onto land, but they rarely travel very far from the water. When threatened by predators, the frogs leap quickly back into the water.

▲ Wallace's flying frog glides between trees using its huge webbed feet like wings.

Frogs and toads that spend more of their time on land, such as the European toad, have a slightly different body. Their heads and bodies are more rounded, their hind limbs are short, and they do not have webbed toes. They may have drier skin.

Jumping

The long legs of the frog allow it to push off from the ground with force and to cover a long distance in one hop. It lands on its front limbs, which cushion the impact. Frogs tend to hop over the ground rather than make long leaps. A leap is a long jump that is many times the frog's body length. Frogs tend to leap only when they are threatened by predators and need to escape quickly.

▼ To leap, a frog quickly unfolds its hind legs to produce a force that propels it forward.

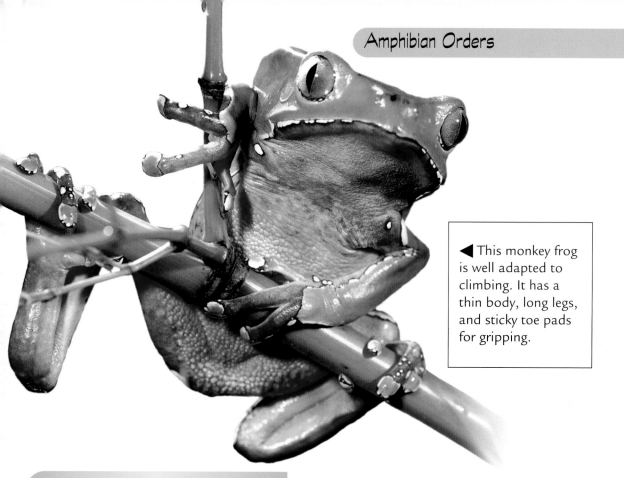

◀ This monkey frog is well adapted to climbing. It has a thin body, long legs, and sticky toe pads for gripping.

Amazing facts

- The Australian rocket frog covers an average of 25 body lengths in a single leap.

- The African pig-nosed frog does not use its legs to burrow. It uses its pointed, strong nose to dig holes.

- When attacked, the common toad swells in size by gulping in air and stands up to make itself look much larger than it really is.

Climbing trees

Tree frogs are well adapted to living in trees. They have thin bodies with long hind limbs. Their toes end in sticky pads that help them to grip the branches, even when they are wet. Tree frogs have good eyesight and can leap from branch to branch. Some even glide through the air. They have large webbed areas between their toes that can be used as parachutes to slow their fall as they leap from tree to tree.

Burrowing

Many frogs and toads make small burrows in which to hide during the day. Frogs living in extreme environments may stay in their burrows for many months to avoid hot or cold weather. Frogs usually burrow backward. They press their heels on the ground and push back and to the side to sweep away the soil. This means that the frog's head is facing forward all the time so that it can watch out for predators.

Parental care

Most frogs and toads lay large numbers of eggs but do not look after their eggs or their tadpoles. Only a few eggs and tadpoles survive to adulthood. The rest are eaten by predators or die from disease.

Some frogs protect their eggs so that they are not eaten before they hatch. One way is to keep the eggs out of water, preventing them from being eaten by fish and other aquatic animals. Some tree frogs build nests of foam. For example, the foam nest frogs of southern Africa gather on tree branches above pools of water to breed. When they breed, the female produces a secretion that the males whip up into a foam using their legs. The female lays her eggs in the foam, which hardens on the outside but remains soft on the inside to protect the eggs. When the tadpoles hatch, the foam dissolves and they fall into the water below.

Amazing facts

- The Australian gastric brooding frog swallows her eggs and they develop in her stomach. She vomits up the tadpoles. These amazing frogs are so rare that they have not been seen since 1985 and may be extinct.

- The female mountain marsupial frog has pouches on her back in which her young develop. The developing young are joined to the mother's blood system so that she can supply them with food.

▼ The male olive midwife toad carries eggs around with him for several weeks, keeping them moist by visiting ponds regularly.

Caring for their eggs

Many tree frogs lay their eggs in the tiny pools of water trapped in the leaves of plants. The female visits the pools of the tadpoles after they have hatched and lays unfertilized eggs, upon which the tadpoles can feed.

▲ Many tropical frogs produce a foam nest in which they lay their eggs.

Some frogs and toads carry their eggs around with them. The midwife toad lays strings of large eggs that are fertilized by the male. The male then wraps these strings around his legs and carries them around. When the eggs are ready to hatch, the male returns to a pool of water and releases the tadpoles. Other species carry their eggs on their backs or in special pouches. For example, the female Surinam toad has special pockets on her back and the male places an egg in each pocket. The young grow inside these pockets and emerge as miniature frogs.

Darwin's frogs have an unusual way of looking after their eggs. These frogs are mouth brooders. The male takes the eggs into the vocal sac under his mouth. The eggs hatch, and the tadpoles remain in the vocal sac. The sac gets steadily larger as the tadpoles grow. They stay inside the vocal sac until they are small frogs and can be released.

▶ Tiny toads are emerging from pockets on the back of this female Surinam toad.

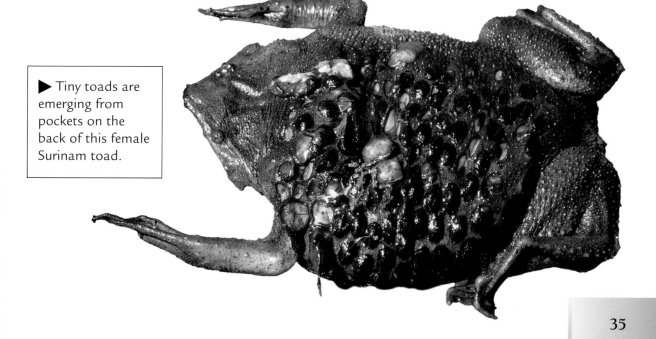

Poison Frogs

Poison frogs are found in the rain forests of Central and South America. These frogs measure from less than 0.5 to more than 2 inches (1.2 to 6 centimeters) long. They secrete poison from their skin. Sometimes they are called poison arrow or poison dart frogs.

Poison frogs are colorful, ranging from striking yellow and bright red to deep blue, orange, and green. Some have spots or flecks, while others have stripes or swirls. These bright colors warn other animals that the frog is poisonous and unsafe to eat. Poison frogs are not eaten by many other animals, but they will eat almost any insect that comes close.

Classification key

CLASS	Amphibia
ORDER	Anura
FAMILY	Dendrobatidae
GENUS	***Dendrobates***
SPECIES	about 170

Poisons

The different species of poison frogs have slightly different poisons, some of which are more harmful than others. Native people living in rain forests have long used frog poisons on the tips of their arrows. When the arrow punctures the skin of an animal, the poison enters their bloodstream and causes paralysis.

Poison frog life cycle

The blue poison frog has an interesting courtship. First, the male and female play by jumping around, chasing, and wrestling with each other. Then, the female lays her eggs in water and the male fertilizes them. After the eggs hatch, the male takes care of the tadpoles until they are old enough to look after themselves. It takes about three months for a blue poison tadpole to metamorphose into a frog.

Some other poison frogs lay eggs on leaves. About four days later the eggs begin to hatch and the tadpoles swim up the male's legs onto his back. He takes them to a safe pool where they are released into the water.

▲ The bright color of the blue poison frog is a clear warning to other animals that this frog is very poisonous.

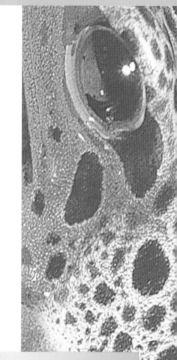

Fighting intruders

Both the male and female poison frogs protect their territory, or space on the forest floor, against other frogs. Frogs can be quite aggressive. First they call out a warning. If this does not work, they try to scare away intruders by chasing them. If all else fails, they fight or wrestle the intruder. Fights usually occur between frogs of the same sex, but sometimes males and females wrestle with one another.

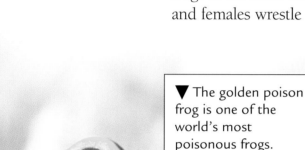

▲ This male poison frog is carrying one of his tadpoles to water.

▼ The golden poison frog is one of the world's most poisonous frogs.

Amazing facts

- The most lethal frog poison comes from the Koikoi poison frog of Colombia. Just 3 ounces (90 milliliters) is enough to kill an adult human being.

- Sometimes a male tricks a female into laying eggs. Instead of fertilizing them, he takes the eggs to feed to his own tadpoles.

Amphibians Under Threat

A mphibians around the world are under threat. There are many reasons for the decline in amphibian numbers, and most are environmental.

Polluted habitats

One of the main threats to amphibians is the destruction of their habitats by people. For example, people pollute streams, rivers, and lakes with chemicals from industry, sewage from homes, and pesticides from agriculture. Amphibians living in these polluted waters cannot survive.

There are many reports of deformed frogs. For example, in 1995 a number of frogs with deformed back legs were found in Minnesota. Similarly deformed frogs have been found in Canada and the United Kingdom. The reasons are uncertain, but such deformities could be caused by pesticides or they could be the result of infection by parasites.

Another reason for the decline in amphibians could be the loss of ozone in the atmosphere. Ozone filters out the harmful ultraviolet light in sunlight. Over the years, certain chemicals have been breaking down the ozone in Earth's atmosphere. This means that more ultraviolet light reaches Earth's surface. Ultraviolet light can damage the skin of people and cause skin cancers. The eggs and larvae of amphibians are also harmed by too much ultraviolet light. This causes them to develop abnormally.

▼ Ponds are essential for frogs and toads. But unless ponds are protected, they may soon fill with silt.

▲ The golden-striped salamander of Spain and Portugal is listed as endangered.

Spreading disease

Amphibians around the world, especially frogs, are being attacked by a disease caused by parasitic fungi. The fungus invades the amphibians' skin and eventually kills them. There have been dramatic declines in the numbers of amphibians in Central America, Australia, and the western United States. Fungal spores are easily spread by animals and by people who pick them up on their clothes.

Amazing facts

- During the last 20 years, at least 10 species of amphibian have become extinct in Australia, New Zealand, and the surrounding islands.
- Worldwide, 91 amphibian species are extinct, missing, or critically endangered. In addition, 193 species are considered endangered or vulnerable.

Alien species

Sometimes amphibians are threatened by other species that are introduced into the habitat. They would not occur naturally. These so-called alien species are often larger amphibians that compete with the native species for food and space. For example, in North America, the red-legged frog of the Pacific coast was once very common. But this species is under threat from large bullfrogs people introduced into their ponds. Similarly, in the United Kingdom, American bullfrogs introduced into backyard ponds compete with native species.

▲ The golden toad, last seen in 1989 at the Monteverde cloud forest reserve in Costa Rica, is now probably extinct.

Protecting Amphibians

One of the most important ways of saving amphibians is to look after their existing habitats and establish new ones.

Many frog, toad, and newt species live in gardens and parks, so building wildlife ponds can help to provide new habitats. An ideal pond has sloping sides so the amphibians can get into and out of the water. Frogs and toads are often called the gardener's friend because they eat many garden pests. On farmland, ponds can be built in the corners of fields. A healthy amphibian population helps to limit the numbers of insect pests.

Nature reserves

Important amphibian habitats can be protected by making them nature reserves. For example, a reserve was set up in Monteverde, Costa Rica for the golden toad. Although these toads have not been seen for some time, the reserve protects many other amphibian species. In the United Kingdom, the great crested newt and the natterjack toad are protected species. No one may collect or kill them, and their breeding sites are protected.

Toad crossings

Each year, thousands of migrating frogs and toads are squashed on the roads as they return to breeding ponds. Many conservation groups have set up toad crossing patrols during the breeding season in which volunteers carry the animals across the roads. Road signs can be placed near the ponds to warn motorists that there may be frogs and toads on the road. Some new road programs have introduced amphibian tunnels that allow the animals to reach the ponds without crossing the road.

▲ In the United Kingdom, signs warn drivers that toads may be crossing the road.

▲ Toad tunnels allow toads to cross underneath major roads safely.

Amazing facts

- In California, plans for a the construction of a new tunnel include moving 100 red-legged frogs, an endangered species living near the site, to a new pond before work begins in 2005.

- The skin of the White's tree frog has been found to have medicinal value. It secretes compounds that act against bacteria and viruses. It also produces a substance that has been used to treat high blood pressure in humans.

Toad barriers

The cane toad was introduced into Queensland in Australia to control some of the pests found in the sugarcane fields. Unfortunately, this large toad has bred so successfully that it has moved into habitats where it threatens the survival of native amphibians. Conservationists have built a barrier to prevent cane toads from spreading into northern Queensland, where it could threaten even more amphibians. Local people are encouraged to collect cane toads so they can be destroyed.

▼ Cane toads are a major problem in Australia, where barriers have been built to keep them out of conservation areas.

Classification

Scientists know of about two million different kinds of animals. With so many species, it is important that they be classified into groups so that they can be described more accurately. The groups show how living organisms are related through evolution and where they belong in the natural world. A scientist identifies an animal by looking at features such as the number of legs or the type of teeth. Animals that share the same characteristics belong to the same species. Scientists place species with similar characteristics in the same genus. The genera are grouped together in families, which in turn are grouped into orders, and orders are grouped into classes. Classes are grouped together in phyla and finally, phyla are grouped into kingdoms. Kingdoms are the largest groups. There are five kingdoms: monerans (bacteria), protists (single-celled organisms), fungi, plants, and animals.

Naming an animal

Each species has a unique Latin name that consists of two words. The first word is the name of the genus to which the organism belongs. The second is the name of its species. For example, the Latin name of the American toad is *Bufo americanus*, and that of the common toad is *Bufo bufo*. This tells us that these animals are grouped in the same genus but are different species. Many animals are given common names, but this may vary from one part of the world to another. For example, the frog *Rhinoderma darwinii* is called Darwin's frog, but is also known by the name of mouth-brooding frog.

For amphibians, a further grouping, the superfamily, is used to show that a group of families have common features but are not different enough from the other families to form a new order.

◀ Many tree frogs have several common names, so a unique Latin name is important.

This table shows how a common American frog is classified.

Classification	Example: common European frog	Features
Kingdom	Animalia	Frogs belong to the kingdom Animalia because they have many cells, need to eat food, and are formed from a fertilized egg.
Phylum	Chordata	An animal from the phylum Chordata has a strengthening rod called a notochord running down its back, and gill pouches.
Subphylum	Vertebrata	Animals that have a backbone, a series of small bones running down the back, enclosing the spinal chord. The backbone replaces the notochord.
Class	Amphibia	Amphibians live in water and on land, and have a smooth, moist skin.
Order	Anura	Anura are amphibians with no tails, a short, rigid back, and long hind legs.
Family	Ranidae	Members of the Ranidae family are egg-laying frogs with long muscular legs and streamlined bodies.
Genus	*Rana*	A genus is a group of species that are more closely related to one another than any group in the family. *Rana* refers to the genus.
Species	*temporaria*	A species is a grouping of individuals that interbreed successfully. The common European frog's species name is *Rana temporaria*.

Amphibian Evolution

▲ Scientists can learn a lot about amphibian evolution from fossils. This frog fossil dates back to about 40 million years ago.

The development of vertebrates that lived on land started about 350 million years ago. At about this time, the climate of the world became hot and dry, causing the water in shallow pools and lakes to dry up. Some fish started to crawl out of the water and began breathing air.

Today, some fish, such as eels, mudskippers, and walking catfish, can live out of water for long periods. However, these fish do not have lungs or a strong fin structure to support their weight on land. The fish ancestors of the first land vertebrates must have had both these features. The most likely ancestors of amphibians were fish called crossopterygians, which were related to the coelacanth, a fish with fleshy fins.

First tetrapods

The earliest fossil tetrapods, or vertebrates with 4 legs, were the Labyrinthodontia, large animals up to 3.2 feet (1 meter) long. A typical example was *Ichthyostega,* a cross between a fish and an amphibian. It had legs and could walk on land. It also had lungs.

About 250 million years ago, there were many different types of tetrapods. Many were smaller than the labyrinthodonts. Most of these tetrapods became extinct, but two groups survived. One evolved into amphibians, and the other evolved into reptiles.

◄ Modern amphibians, such as this toad, are thought to have evolved from a group of ancestral amphibians called *Lissamphibia.*

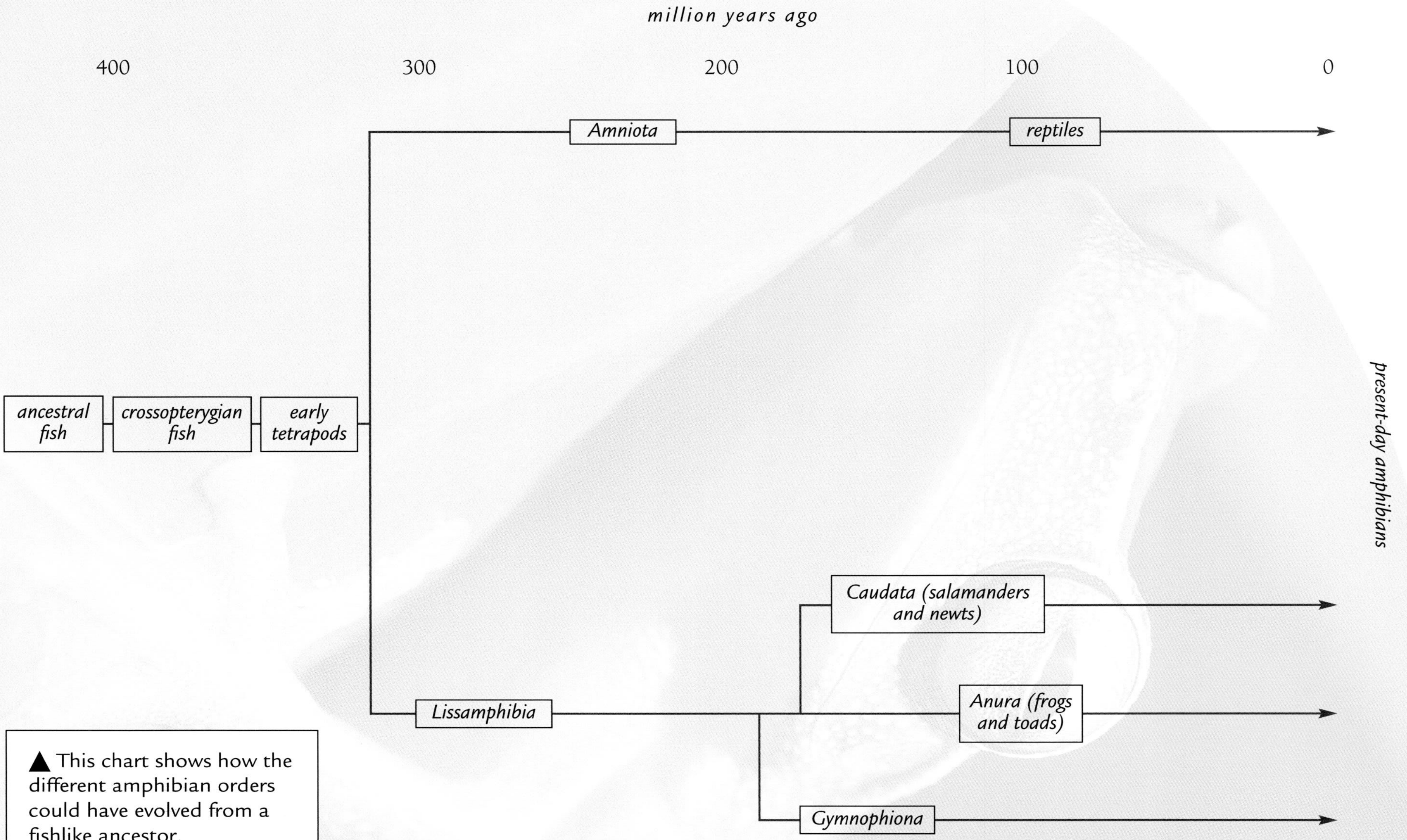

present-day amphibians

million years ago

0

100

200

300

400

reptiles

Amniota

Caudata (salamanders and newts)

Anura (frogs and toads)

Gymnophiona

Lissamphibia

early tetrapods

crossopterygian fish

ancestral fish

▲ This chart shows how the different amphibian orders could have evolved from a fishlike ancestor.

Glossary

adapt change in order to cope with the environment

algae simple, nonflowering plants found in water

amphibian animal that lives part of its life on land and part in water, though there are a few exceptions

aquatic living in water

blood vessel tiny tube that transports blood around an animal's body

breed reproduce

caecilian wormlike amphibian

camouflage coloring that blends with the background, making an animal difficult to see

carnivore animal that eats other animals

characteristic feature or quality. For an animal, a characteristic would be having hair or having wings.

clutch more than one egg laid by one female

dormant inactive

ectothermy having a body temperature that rises and falls with the outside temperature, often referred to as cold-blooded

embryo unborn or unhatched young

evolve change very slowly over a long period of time

evolution slow process of change in living organisms so they can adapt to their environment

extinct no longer in existence; to have permanently disappeared

fertilize cause a female to produce young through the introduction of male reproductive material

fossil trace or impression of ancient life preserved in rock

frogspawn mass of frog's eggs

gill part of the body that an aquatic animal uses to collect oxygen from water

gland organ that releases a substance such as saliva or sweat

herbivore animal that eats plants

interbreed mate with another individual of the same species

invertebrate animal that does not have a backbone

larva young animal that looks different from the adult and changes shape as it develops

mammal class of vertebrates that feed their young milk, are usually covered in hair, and have a constant body temperature

mate reproduction partner of the opposite sex; fertilize the eggs of a female of the same species

metamorphosis process of changing from one form to another during development. In amphibians, metamorphosis is usually associated with changing from an aquatic larval stage to a terrestrial adult stage.

migrate make a regular journey, often related to the changes of the seasons

mucus slimy substance released by the skin of amphibians

native local to or born in an area

nocturnal active at night

northern hemisphere half of Earth that lies above the equator

nutrient substance that provides food for an organism

organism living being, such as an animal, plant, or bacterium

paralysis condition in which the muscles cannot move

parasite animal that lives on or in another animal

pigment natural skin coloring

predator animal that hunts other animals

prey animal that is hunted by other animals

reptile cold-blooded, egg-laying vertebrate with tough skin covered in scales

secretion substance released from part of the body

southern hemisphere half of Earth that lies below the equator

species group of individuals that share many characteristics and which can interbreed to produce offspring

streamlined having a slim shape that moves through water easily

superfamily group of families with similar features that are not different enough from other families to form a separate order

surface area total area of the outside of an organism or object

tadpole larval stage in the life cycle of a frog or toad

terrestrial living on land

territorial describing the behavior of an animal that is defending its territory

territory range or area claimed by an animal or group of animals

torpor state of inactivity

tropics region of the world that lies on either side of the equator, with a hot, often wet climate

vertebrate animal that has a backbone

viviparous giving birth to live young

Further Information

Bailey, Jill. *Frogs and Toads*. Chicago: Raintree, 2004.

Fullick, Ann. *Ecosystems & Environment*. Chicago: Heinemann Library, 2000.

Parker, Edward. *Reptiles and Amphibians*. Chicago: Raintree, 2003.

Sachidhanandam, Uma. *Threatened Habitats*. Chicago: Raintree, 2004.

Schroeder, Patricia. *Reptiles and Amphibians: Scales, Slime, and Salamanders*. Chicago: Raintree, 2000.

Spilsbury, Louise and Richard Spilsbury. *Classifying Amphibians*. Chicago: Heinemann Library, 2003.

Townsend, John. *Incredible Amphibians*. Chicago: Raintree, 2005.

Whyman, Kathryn. *The Animal Kingdom: A Guide to Vertebrate Classification and Biodiversity*. Chicago: Raintree, 2000.

Index